Four-Element Model of Particle Physics

By Jen Chyi Liao

Copyright © 2019 by Jen Chyi Liao

Contact: liao_jen@hotmail.com

Four-Element Model of Particle Physics

Contents

Chapter 1: Introduction 1

Chapter 2: Basic Concepts of Particle Physics 2

    Fundamental Particles 2

    Particle and Antiparticle 4

    Unit of Particle Mass 4

    Electron and Positron 5

    Photon 6

    Neutrino 7

    Muon and Tauon 8

    Quark 8

    Proton and Neutron 8

    Fundamental Forces 9

    Virtual Particles and the Vacuum 10

Chapter 3: Four-Element Model 11

    Postulate 1 12

    Postulate 2 13

    Postulate 3 14

    Postulate 4 15

    Postulate 5 15

    Postulate 6 16

    Postulate 7 16

    Postulate 8 17

Four-Element Model of Particle Physics

Chapter 4: The Four Elements 19
    Electron-positron Pair Creation and Annihilation......... 19
    N-bit and P-bit 21
    Energy Carried by a Bit.. 24
    Motion of Bit.. 24
    Interaction of Bit 26
    Positive Charge and Negative Charge............................ 26

Chapter 5: Particle Structures and Properties 30
    Electron and Positron 30
    Electron-neutrino and Anti-electron-neutrino................ 32
    Particle Spin 33
    Photon... 34
    Photon Polarization 35
    Photon and Electromagnetic Wave 36

Chapter 6: Electromagnetic Force and Radiation 38
    Electric Force 38
    Magnetic Force... 41
    Emission of Bits 42
    Electromagnetic Radiation 43

Chapter 7: Origin of Mass and Gravity............................... 48
    Origin of Mass.. 48
    Heavy Particles... 49
    Muon and Tauon 50
    Muon-neutrino and Tauon-neutrino 51

… Origin of Gravity … 51

Chapter 8: Interaction between Photon and Atom … 53
    Interaction between Electron and Atomic Nucleus … 53
    Interaction between Photon and Atom … 55
    Photoelectric Effect … 57
    Compton Effect … 58
    X-ray Production … 59

Chapter 9: Quark Mystery … 61
    Color Charge … 61
    Quark Confinement and Hadronization … 62

Chapter 10: Miscellaneous Topics … 64
    Wave-particle duality … 64
    Energy and Momentum Conservation … 71
    Neutrino Mass … 74

References … 75

Four-Element Model of Particle Physics

Chapter 1: Introduction

Ancient Greek philosophers introduced a "Four Element Theory" saying that the four most basic elements of our universe are earth, water, air and fire. From modern view of particle physics, the most basic elements of our universe are the so-called fundamental particles such as electrons and quarks. In this book, I am going to present a "Four-element Model" with a novel idea that the most basic elements of our universe are: positive charge, negative charge and two tiny energy carriers named p-bit and n-bit.

The underlying concept of the four-element model is that everything in our universe is made from the above-mentioned four basic elements. All forces between particles are the result of interactions among these elements. Mass and gravity have the same origin based on this concept.

Chapter 2: Basic Concepts of Particle Physics

Before venturing into the world of the four-element model, let us review briefly some basic concepts of particle physics.

The best current theory of particle physics is the so-called standard model. Contents of this chapter are mainly based on this model.

Fundamental Particles

Fundamental particles are the basic building blocks of our universe. They are point-like in the sense that they don't have internal structure, and cannot be divided into smaller constituents. Fundamental particles and their properties are shown in Table 2.1. Some particles, such as the W⁻ particle, are left out deliberately as they are irrelevant to the main subject of this book.

Four-Element Model of Particle Physics

Table 2.1 Fundamental particles

Name	Symbol	Electric charge	Spin	Mass MeV
electron	e⁻	-1	1/2	0.511
muon	μ	-1	1/2	105.7
tauon	τ	-1	1/2	1,784
electron-neutrino	v_e	0	1/2	Small, but not 0
muon-neutrino	v_μ	0	1/2	< 0.27 eV
tauon-neutrino	v_τ	0	1/2	< 31 eV
photon	γ	0	1	0
up quark	u	+2/3	1/2	5
down quark	d	-1/3	1/2	7
charm quark	c	+2/3	1/2	1,500
strange quark	s	-1/3	1/2	150
top quark	t	+2/3	1/2	170,000
bottom quark	b	-1/3	1/2	5,000

Particles are characterized by their mass, electric charge, and spin. Spin is the intrinsic angular momentum of a particle and is measured in units of the Planck constant, h divided by 2π. A particle of spin ½ has an intrinsic angular momentum of the amount of $h/4\pi$.

Particle and Antiparticle

Every particle has an associated antiparticle. Some particles are identical to their antiparticles and photon is one of them. An antiparticle has the same mass as its associated particle but has an electric charge of the opposite sign. For example, the electron has an electric charge of -1, and its antiparticle, the positron, has an electric charge of +1. When a particle and its antiparticle meet, they annihilate each other and become energy, typically in the form of photons.

In most literatures, some antiparticles are denoted by writing a bar over their particle names or symbols. For example, the symbol for the proton is p, and the symbol for the antiproton is $\bar{p}$. For most charged particles, a superscript - is used to denote a particle and a superscript + is used to denote the associated antiparticle. For example, the symbol for the electron is e^-, and the symbol for the positron is e^+.

Unit of Particle Mass

According to Albert Einstein's Special Theory of Relativity, mass and energy are equivalent and are related by

$$E = mc^2$$

or

$$m = \frac{E}{c^2}$$

where c is the speed of light. Physicists found that it is more convenient to use energy unit to specify the mass of a particle, and the most commonly used energy unit is electron-Volt (abbreviated as eV). Thus, "eV/c²" is used as the unit for particle mass and commonly with "c²" dropped for succinctness. The unit electron-Volt (eV) is related to Kg by

$$1\frac{eV}{c^2} = 1.78 \times 10^{-36} Kg$$

The mass of an electron is 9.11 x 10⁻³¹ Kg which is equivalent to 0.511 MeV/c² (511,000 eV/c²), or simply 0.511 MeV.

Electron and Positron

Some particles have mass and some don't. Particles with mass are called matter particles. The electron and positron are the lightest matter particles. The electron carries a negative charge of one unit, and the positron carries a positive charge of one unit.

The electron and positron are antiparticle to each other. When an electron and a positron meet, they annihilate each other and generate two or more photons as shown in Fig. 2.1.

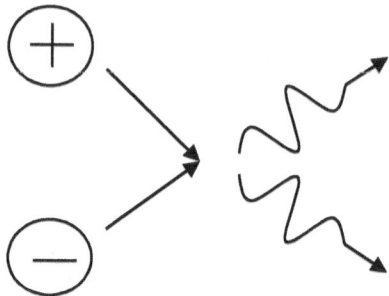

Fig. 2.1 Electron-positron pair annihilation

Photon

The photon has no mass. It interacts with charged particles but not with other photons. The photon is its own antiparticle.

The energy of a photon can be converted into particle masses. In a process called electron-positron pair creation, a photon is converted into an electron and a positron as shown in Fig. 2.2.

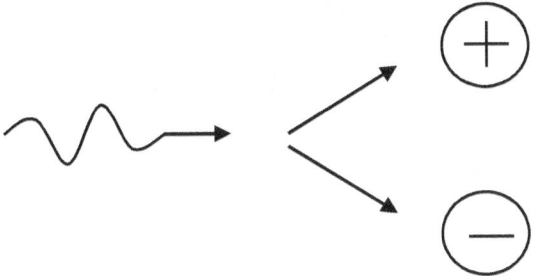

Fig. 2.2 Electron-positron pair creation

Neutrino

There are three different types of neutrinos: electron-neutrino, muon-neutrino, and tauon-neutrino. The mass of all types of neutrinos shown in Table 2.1 are estimated values because they are too small to be measured accurately. Since the three types of neutrinos differ only in their masses, it is hard to differentiate one from another. However, in particle interactions, an electron-neutrino is always produced in association with an electron, and is never with a muon or a tauon. Similarly, a muon-neutrino is always produced in association with a muon, and a tauon-neutrino is always produced in association with a tauon.

Neutrinos have no electric charge and seldom interact with other particles. The vast majority of neutrinos detected on Earth come from the Sun.

Muon and Tauon

The muon and tauon are unstable. The muon decays into a muon-neutrino, an electron, and an anti-electron-neutrino. The process can be expressed as

$$\mu \to e^- + \bar{v}_e + v_\mu$$

The decay of a tauon is the same as that of a muon except that the final products include a tauon-neutrino instead of a muon-neutrino.

Quark

Quarks are ingredients of particles called hadrons. Hadrons are grouped into two categories: baryons and mesons. A baryon is formed by three quarks, and a meson is formed by a quark and an anti-quark. All hadrons, except the proton and neutron, are unstable. The neutron is stable only inside a nucleus. A free neutron decays quickly into a proton.

Quarks exist only inside hadrons and cannot be observed as free quarks.

Proton and Neutron

The proton and neutron are baryons. They are ingredients of the atomic nucleus, and are thus called the nucleons. The proton carries a positive electric charge of one unit, and with a mass of 938.28 MeV, about 2,000 times of that of an electron. The neutron is electric charge neutral, and with a mass of 939.573 MeV, only a little bit larger than that of a proton.

The proton and neutron were once considered as fundamental particles, but now physicists know that they are made of quarks.

Fundamental Forces

There are four fundamental forces acting among particles: the gravitational force, the electromagnetic force, the strong force, and the weak force.

Interactions due to the electromagnetic force are called electromagnetic interactions. Similarly, interactions due to the strong force and the weak force are called strong interactions and weak interactions, respectively.

The gravitational force, also known as gravity, is a long range force acting upon all matter particles, and is always attractive. The gravitational force is very weak, but as it is always attractive and as matter objects are made of huge number of matter particles, the force between large matter objects can become extremely large. An example is the gravity between planets.

The electromagnetic force is about forty times stronger than the gravitational force. It exists only between particles with electric charge, and is repulsive between particles with charges of the same sign and attractive between particles with charges of the opposite sign.

The strong force is the strongest among the four fundamental forces. It exists only inside the nucleus, and is the force that binds quarks together to form hadrons. It is also the force that binds nucleons together to form a nucleus.

Like the strong force, the weak force exists only inside the nucleus. It can transform a particle from one type to another.

Virtual Particles and the Vacuum

All particles that have been discussed so far are real particles in the sense that they can be observed in the laboratory. There are particles that cannot be observed in laboratory experiments. Particles of this type are called virtual particles.

According to Werner Heisenberg's principle of uncertainty, a virtual particle of an energy E can exist for a time duration T as long as the following condition is satisfied:

$$ET \leq h$$

The above expression means that the product of E and T is less than or equal to the Planck constant h.

From the modern view of particle physics, in the vacuum pairs of virtual particle and antiparticle are constantly created. The vacuum is not empty at all, and is instead like a sea of virtual particles.

Chapter 3: Four-Element Model

According to the Big Bang Theory, our universe started from the big bang of a tiny point. The big bang resulted into a fire ball of extremely high temperature. At that time, there is energy only. Very soon energy began to convert into fundamental particles. Fundamental particles are then bound to form nucleons which in turn clustered to form nucleus. Finally, a nucleus together with one or more electrons forms an atom.

Before the fundamental particles are created, there is energy only. We also know that when a particle and an antiparticle meet, they annihilate each other and become energy. What is energy and how is it converted into matter? To answer this question, I have developed a set of postulates and call them collectively the four-element model.

The main idea of the four-element model is that all particles, including fundamental particles, are made from four basic elements, and all forces, including the four fundamental forces of the standard model, are the results of interactions among these four elements.

The four-element model consists of eight postulates.

Postulate 1

The four elements of the universe are: positive charge, negative charge, n-bit and p-bit. The n-bit and p-bit are energy carriers of the universe. The n-bit interacts only with the negative charge, and the p-bit interacts only with the positive charge. The negative charge does not interact with the positive charge. The n-bit does not interact with the p-bit.

The n-bit and p-bit are so named because they are tiny and each carries a tiny amount of energy. The negative charge and positive charge of the four elements postulated here are not the same as those in conventional physics.

From now on, "bit" will be used as the common name of the n-bit and p-bit, and "charge" will be used as the common name of the positive charge and negative charge.

Shown in Fig. 3.1 are symbols to be used for the four elements. There are two symbols for each type of bit, one with its direction shown and one without.

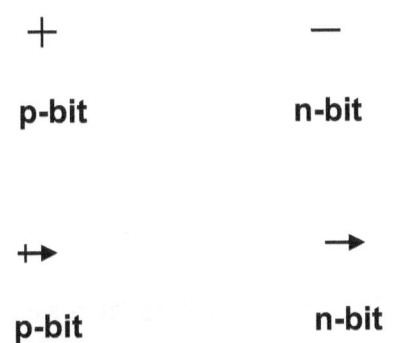

Fig. 3.1a Symbols for the bit

Fig. 3.1b Symbols for the charge

Postulate 2

An electron-neutrino is formed by a group of n-bits, and an anti-electron-neutrino is formed by a group of p-bits, as shown in Fig. 3.2.

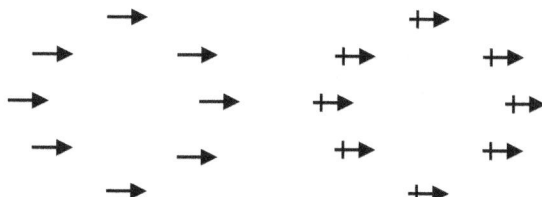

Fig. 3.2 Electron-neutrino and anti-electron-neutrino

Postulate 3

A photon is a composite of an electron-neutrino and an anti-electron-neutrino, as shown in Fig. 3.3.

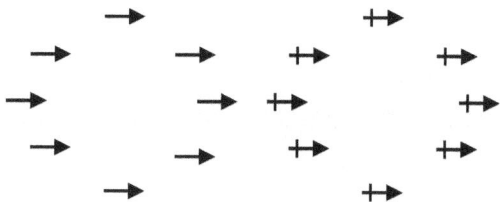

Fig. 3.3 Photon

Postulate 4

An electron contains a negative charge and an electron-neutrino. A positron contains a positive charge and an anti-electron-neutrino.

Postulate 5

There are bits and charge pairs in the vacuum. A charge pair consists of a positive charge and a negative charge as shown in Fig. 3.4. The charge pair is electric neutral because the two charges are close to each other.

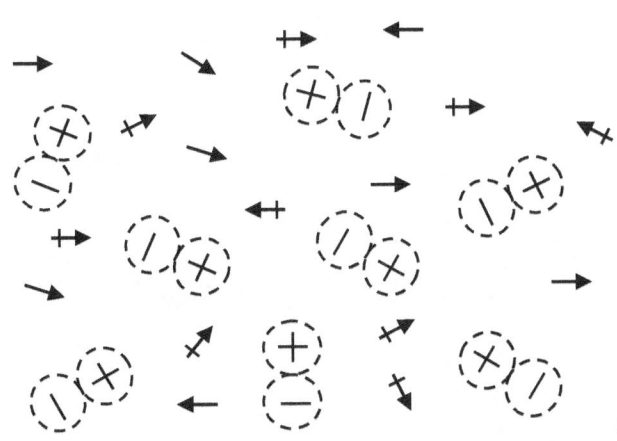

Fig. 3.4 Bits and charge pairs in the vacuum

Postulate 6

Mass is a joint effect of charges and bits.

When a group of bits are attracted by charges to form a matter particle, as shown in Fig. 3.5, the energy of the bits becomes the mass of the particle.

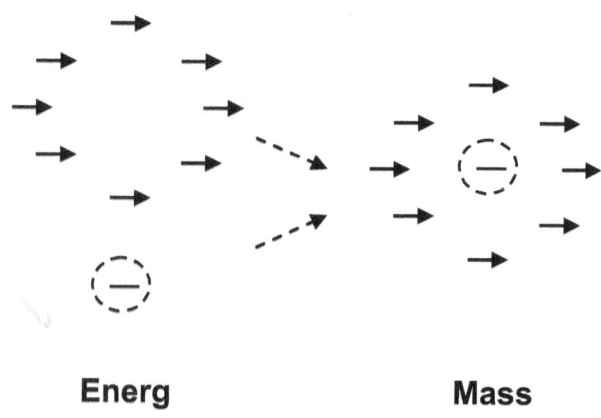

Energ   **Mass**

Fig. 3.5 Energy to mass

Postulate 7

All forces are the results of interactions among the four basic elements.

The four fundamental forces described in the last chapter are manifestation of the same force under different circumstances. When the separation of the two particles is extremely small, say about or below the size of a nucleus, the force manifests as the strong force. When the separation is large, the force manifests as the electromagnetic and gravitational forces. When it comes to some particle transformations or decays, the force is called the weak force by particle physicists.

Postulate 8

The n-bit is the anti-element of the p-bit, and vice versa. The negative charge is the anti-element of the positive charge, and vice versa. Replacing every constituent element with its anti-element changes a particle into its antiparticle.

The electron consists of n-bits and a negative charge, and the positron consists of p-bits and a positive charge, as shown in Fig. 3.6. If the negative charge is replaced with a positive charge and each n-bit replaced with p-bit, the electron becomes a positron. Conversely, if the positive charge is replaced with a negative charge and each p-bit replaced with n-bit, the positron becomes an electron.

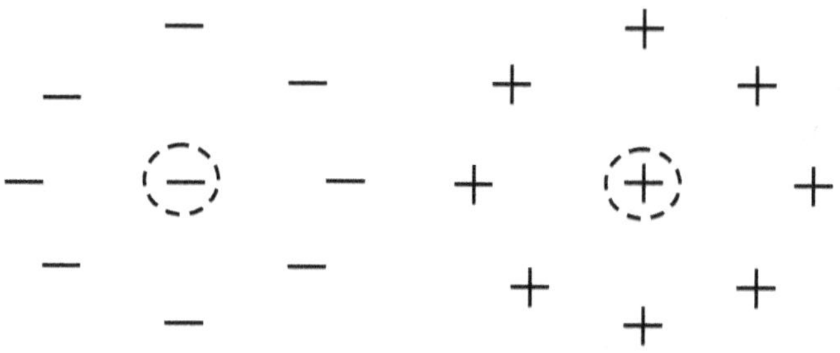

Four-Element Model of Particle Physics

Fig. 3.6 Anti-element and antiparticle

Chapter 4: The Four Elements

Why are the four elements postulated as the basic elements of the universe? What are the characteristics of these elements? The answers are provided in this chapter.

Electron-positron Pair Creation and Annihilation

When an electron and a positron meet, they annihilate each other and create two or more photons as shown in Fig. 2.1. As the photon does not carry electric charge, what happen to the charges of the electron and the positron? Is it possible that each photon carries some equal amount of positive and negative charges and is thus electric neutral? But as far as we know the electric charge of an electron or a positron cannot be divided. I postulated that the electron and positron are both a composite of an electric charge and something else; and that in the process of pair annihilation, the positive charge of the positron and the negative charge of the electron stick closely together as a pair. The charge pair as a whole is charge neutral and does not interact with any charged particles. The positive and negative charges are in a sense disappear into the background of the universe.

What are the "something else" in an electron and a positron? I postulated that the one in the electron is an electron-neutrino, and the one in the positron is an anti-electron-neutrino. I also postulated that the photon is a composite of an electron-neutrino and an anti-electron-neutrino. In the process of electron-positron pair creation, the electron-neutrino of the photon combines with the negative charge to form an electron, and the anti-electron-neutrino of the photon combines with the positive charge to form a positron, as shown in Fig. 4.1.

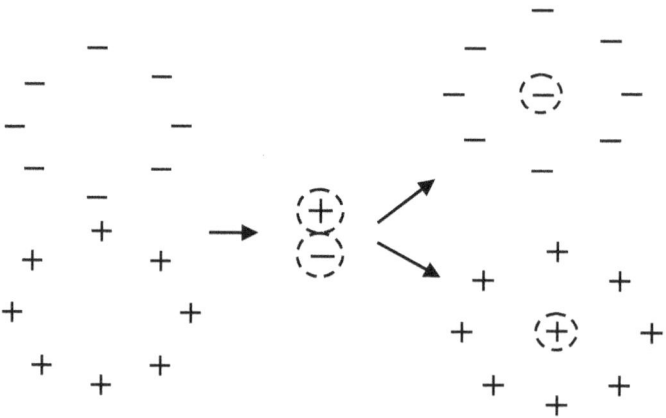

Fig. 4.1 Electron-positron pair creation

N-bit and P-bit

In modern view of particle physics, a particle has wave property, and a wave has particle property. This strange feature is called wave-particle duality. The wave and particle properties are related by the following expressions

$$E = h\nu$$

$$p = \frac{h}{\lambda}$$

In the above expressions, E is energy, h is the Planck constant, ν is frequency, p is momentum, and λ is wavelength.

The value of h is

$$h = 6.6262 \times 10^{-34}\ Js$$

There are three questions about the frequency of a particle:

- What does the frequency means for a particle?
- Why the frequency of a particle is related to the energy of the particle?
- How a particle changes its frequency?

Frequency is always associated with something that oscillates periodically and is thus, in a sense, of time-varying nature. It is indeed so for waves that we are familiar with in our daily life. For example, the frequency of a sound wave is related to the oscillation of amplitude. How about the frequency of a matter particle? Is there something oscillating? If so, is it an up-down oscillation, a rotation around a center, or something else? I found some clue by examining the operation of an electronic device, the vacuum tube.

A vacuum tube has an anode and cathode as shown in Fig. 4.2. The anode and cathode are connected to the positive and negative terminals, respectively, of a DC power supply. The cathode is heated so that electrons can be emitted from the surface. The electrons then fly from the cathode to the anode due to the electric field. The electrons accelerate on their way from the cathode to the anode. The electrons' energies, or equivalently the frequencies, increase constantly when they fly from the cathode to the anode.

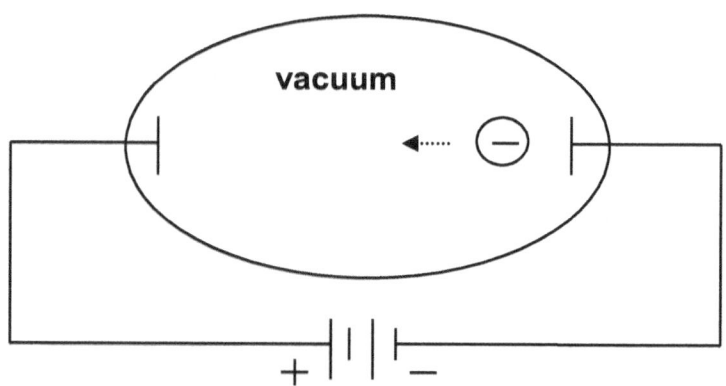

Fig. 4.2 Vacuum tube

The electric field inside the vacuum tube is a static field. How does such a static field change an electron's frequency which is a property of time-varying nature? The same question can be asked differently, i.e., how an electron acquires energy in an electric field?

The questions about the frequency of a particle can be answered by positing that there exist some tiny things in the electric field, and that an electron acquires additional energy by absorbing some of these tiny things. The tiny thing is named "bit" as it is a tiny bit carrying an extremely small amount of energy.

There are two kinds of bit: n-bit and p-bit. When an electron flies from the cathode to the anode, some n-bits join and become a part of the electron. The electron thus gains additional energy, and its frequency is increased.

The postulate of bit provides the following answers to the questions about particle frequency:

- The frequency of a particle is simply a measure of the energy that the particle possesses, and needs not be associated to something that is oscillating.
- The frequency, or equivalently the energy, of a particle is in proportion to the number of bits the particle contains.
- A particle's frequency increases when the particle acquires additional bits.

The n-bit is attracted by the negative charge of an electron and becomes a part of the electron. Since the electron consists of a negative charge and an electron-neutrino, the electron-neutrino is a group of n-bits.

The p-bit is the anti-element of the n-bit. It is attracted by the positive charge of a positron and becomes a part of the positron. Since the positron consists of a positive charge and an anti-electron-neutrino, the anti-electron-neutrino is a group of p-bits.

Energy Carried by a Bit

What is the energy carried by a single bit? The answer can be found from the expression that relates the energy to the frequency of a particle

$$E = h\nu$$

Since ν is in direct proportion to the number of bits contained in the particle, the energy carried by a single bit is kh joule and k is a constant. The value of k is unknown as the ratio of ν to the number of bits is unknown. Since h, the Planck constant, is extremely small; the energy carried by a bit is extremely small, and all known particles contain a huge number of bits.

Motion of Bit

Since the neutrino is formed by a group of bits and the neutrino travels with the speed of light, the bit must travel with the speed of light.

Some particle has spin angular momentum, or spin for short. A particle with its spin anti-aligned with its travel direction is called left-handed, and a particle with its spin aligned with its travel direction is called right-handed.

Since the spin of a particle must come from its constituent bits, the bit has spin. The spin of a bit is due to the rotation around the line along the travel direction.

Since the electron-neutrino is always left-handed, the n-bit must be left-handed. Similarly, since the anti-electron-neutrino is always right-handed, the p-bit must be right-handed. The spin and travel directions of n-bit and p-bit are shown in Fig. 4.3.

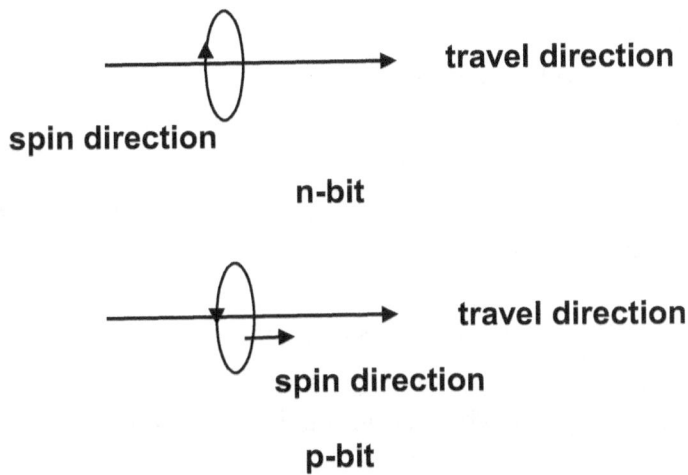

Fig. 4.3 Motion of bit

Interaction of Bit

A bit does not interact with any other bit. It is so because otherwise a photon will interact with other photons as the photon is formed by a group of n-bits and a group of p-bits.

The n-bit interacts only with the negative charge, and the p-bit interacts only with the positive charge.

Positive Charge and Negative Charge

The charge has the following properties:

- It can attract bits to form a matter particle such as an electron and positron.
- Its effect is of infinite range.
- It can be carried along by bits.
- When a negative charge and a positive charge stick closely together, they become a charge-neutral pair.

Judging from its properties, a charge has two parts: a charge center and many charge lines of infinite length, as shown in Fig. 4.4. The charge line is not an intrinsic part of a charge.

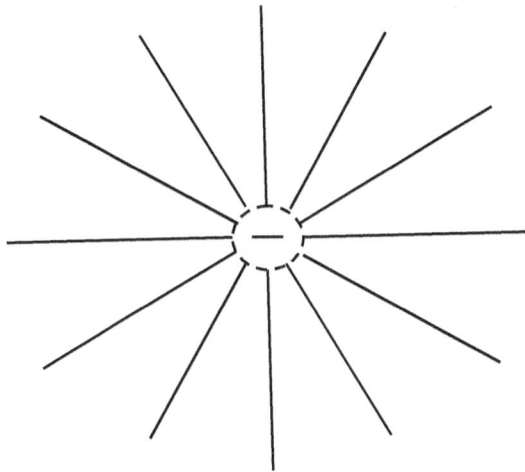

Fig. 4.4 Charge center and charge lines

There are a lot of charge lines uniformly distributed in the universe as shown in Fig. 4.5. When a charge line is in contact with the charge center of a charge, it becomes a part of that charge. A charge line can be in contact with many charge centers, positive or negative. The sign and strength of a charge line thus depend on the types, positive or negative, and the number of charge centers that the charge line contacts.

Four-Element Model of Particle Physics

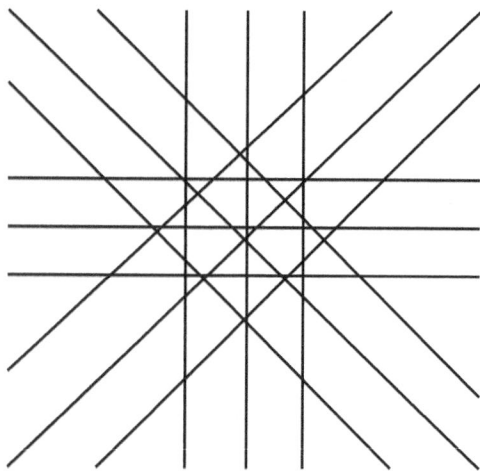

Fig. 4.5 Universal charge lines

Why the charge line is not an intrinsic part of a charge? Because if the charge line is an intrinsic part of a charge; when a charge moves, a lot of charge lines, which are of infinite length, will move with the charge center, and will likely to tangle with charge lines of other charges.

According to classical theory of electromagnetism, like charges repel and opposite charges attract, and the force falls as the square of the distance between the charges. Do charges of the four-element model behave in the same way? The answer is 'no' for the argument below.

If charges of the opposite sign attract in the same way as conventional electric charges; the force between charges of a charge pair would become extremely strong, and positron-electron pair creation is unlikely to happen. Therefore, there is no mutual attraction or repulsion among charges.

A positive charge interacts only with p-bits, not with n-bits. Similarly, a negative charge interacts only with n-bits, not with p-bits. The electromagnetic force between two charged particles is the result of interactions between constituent charges and bits of the two particles. When two charged particles are very close to each other, the intensity of the electromagnetic force does not approach infinity as is predicted by classical theory of electromagnetism.

Chapter 5: Particle Structures and Properties

How the four elements combine to form particles? This chapter aims to provide the answer as well as to describe structures and properties of some particles.

Electron and Positron

An electron is formed by a negative charge and a group of n-bits. Most of the bits are located in a small region around the charge center. The linear momentum of the electron is the vector sum of that of the bit. The spin of the electron is also the vector sum of that of the bit. Similarly, a positron is formed by a positive charge and a group of p-bits. The structure of an electron and the structure of a positron are shown in Fig. 5.1.

When the n-bits are replaced with p-bits, and the negative charge with positive charge, an electron is turned into a positron. Conversely, when the p-bits are replaced with n-bits, and the positive charge with negative charge, a positron is turned into an electron. For any particle, replacing every constituent element with its anti-element changes it into its anti-particle.

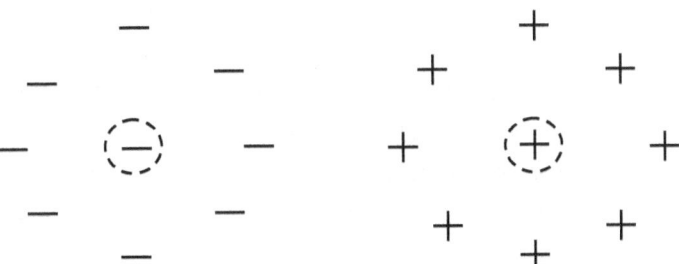

Fig. 5.1 Electron and positron

An electron is at rest because its bits are distributed such that the net linear momentum is zero. To make the electron move in a certain direction, additional bits with net linear momentum in this direction must be added.

The energy of an electron is determined by the number of bits the electron has. When an electron is at rest, its energy and mass are called the rest energy and rest mass of that electron.

We need to be careful with the term "at rest". According to Einstein's special theory of relativity, an object is at rest only with reference to a specific reference frame. An object that is at rest in one reference frame is not at rest in other reference frames.

Electron-neutrino and Anti-electron-neutrino

When the n-bits of an electron leave the electron, they become an electron-neutrino. The electron-neutrino is thus a group of n-bits. The spin and linear momentum originally possessed by the electron become those of the electron-neutrino. Similarly, when the p-bits of a positron leave the positron, they become an anti-electron-neutrino. The anti-electron-neutrino is thus a group of p-bits. The spin and linear momentum originally possessed by the positron become those of the anti-electron-neutrino. The structure of an electron-neutrino and the structure of an anti-electron-neutrino are shown in Fig. 5.2.

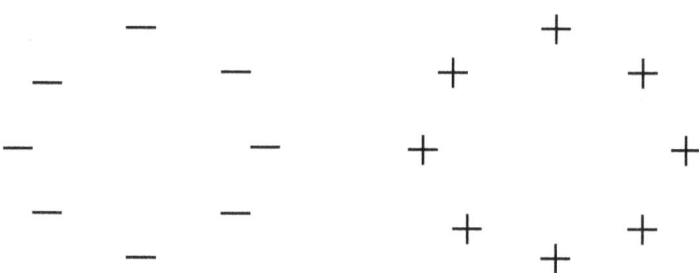

Fig. 5.2 Electron-neutrino and anti-electron-neutrino

The n-bits of an electron are bound to the electron because of the attraction of the negative charge. When the n-bits leave the electron and become an electron-neutrino, they are just a group of unbound entities moving together in the same direction and with the same speed. If one of the bits changes its direction or speed, it will not stay with the group.

Since the n-bit is left-handed and travels with the speed of light, the electron-neutrino is left-handed and travels with the speed of light. Similarly, the anti-electron-neutrino is right-handed and travels with the speed of light.

Particle Spin

Since the spin of a particle must come from the spins of its constituent bits, it seems that the spin of a particle should increase with its energy which is in proportion to the number of bits. But the spin of a particle is a constant, why? A possible explanation is given below.

When a large number of bits gather around a charge center, the angular momentums of two adjacent bits tend to cancel each other, as shown in Fig. 5.3. As a result, only those in the outmost region contribute to the net angular momentum of the particle. Therefore, the spin of a particle is independent of the number of bits, or equivalently the energy, of the particle.

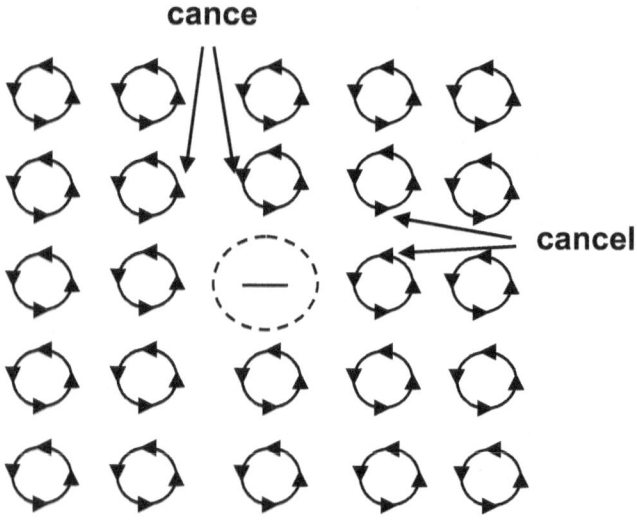

Fig. 5.3 Cancellation of interior angular momentum

The above is my own explanation. The origin and value of particle spin is a mystery in particle physics.

Photon

The photon consists of an electron-neutrino and an anti-electron-neutrino as shown in Fig. 5.4.

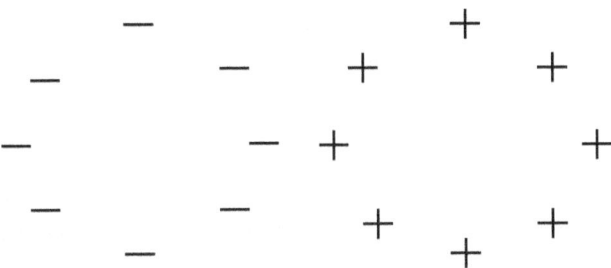

Fig. 5.4 Photon

Since the bit does not interact with other bits, a photon does not interact with other photons.

When the n-bits of a photon are replaced with p-bits, and the p-bits with n-bits; the resultant particle is still a photon. Thus, the photon is its own antiparticle.

Since the electron-neutrino, which is left-handed, and the anti-electron-neutrino, which is right-handed, must travel in the same direction; the vector sum of their spin should be 0. But the photon spin is 1. This is a puzzle to me.

Photon Polarization

Electromagnetic wave has a property called "polarization". Polarization of a wave is described by the locus of the tip of the vector of the electric field as time progresses. If the locus is a straight line, the wave is said to be linearly polarized. If the locus is a circle, the wave is said to be is circularly polarized. Since light is an electromagnetic wave, it has the property of polarization too.

Since light is a collection of photons, the property of polarization implies that the photon has polarity. When we say that a subject has polarity, it implies that this subject can be divided into two parts with properties different in a certain way. The two parts are referred to as poles of this subject. A photon has two parts: the electron-neutrino and the anti-electron-neutrino. The electron-neutrino and the anti-electron-neutrino are thus the two poles of a photon,.

A light wave certainly can be linearly or circularly polarized, but how about an individual photon? Since both the neutrino and antineutrino of a photon must travel with fixed direction and speed in free space, they don't rotate around each other. Thus, the polarity is fixed when a photon travels in free space.

Photon and Electromagnetic Wave

The time variation of a periodic electromagnetic wave at a space point is shown in Fig. 5.5. In this figure, the amplitude changes continuously, and the polarity changes once every period.

Since an electromagnetic wave is a collection of photons; the amplitude of the wave at a space point represents the density of photons at that space point, and the sign of the amplitude reflects the polarity of the photons. In Fig. 5.5, the first half period of the wave represents a group of photons of the same polarity, and the second half a group of photons of polarity opposite to that in the first half.

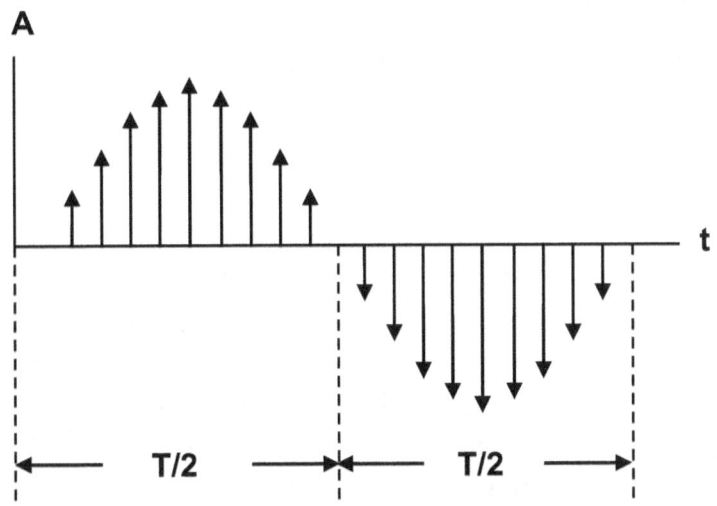

Fig. 5.5 Time variation of a periodic electromagnetic wave

Lights from the Sun and electric lamps are un-polarized meaning that not all the photons at a space point are of the same polarity. An un-polarized light cannot be depicted as a wave like that in Fig. 5.5, and there is no wave frequency associated with it. Nonetheless, the frequency of each individual photon of an un-polarized light can still be determined from the energy of the photon. For example the sunlight consists of a spectrum of photons of different frequencies, from infrared to ultraviolet.

Chapter 6: Electromagnetic Force and Radiation

According to classical theory of electromagnetism, electromagnetic force is due to electric and magnetic fields; but the standard model says that the electromagnetic force is the result of exchanging photons between charged particles. In this chapter, a different view based on the four-element model will be provided.

Electric Force

When a positron or an electron is at rest, their constituent bits are distributed such that the resultant linear momentum is zero as shown in Fig. 6.1. In the figure charge lines are omitted as usual for succinctness.

It should be pointed out that Figure 6.1 is a simplified picture as actual directions and distribution of the bits are much more complicated.

Charge lines of two charged particles always overlap such that charge lines of one particle are shared by or in proximity of another particle, as shown in Fig. 6.2. Overlapping of charge lines can cause redistribution of bits.

Four-Element Model of Particle Physics

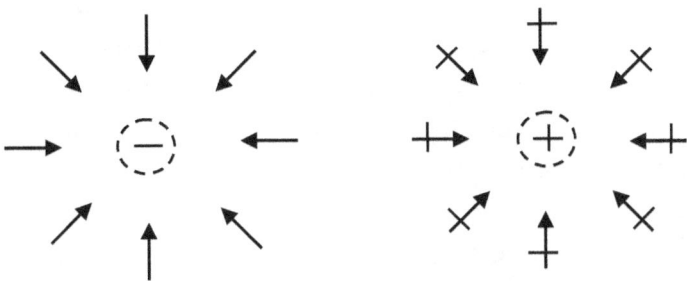

Fig. 6.1 Bit distribution of electron and positron at rest

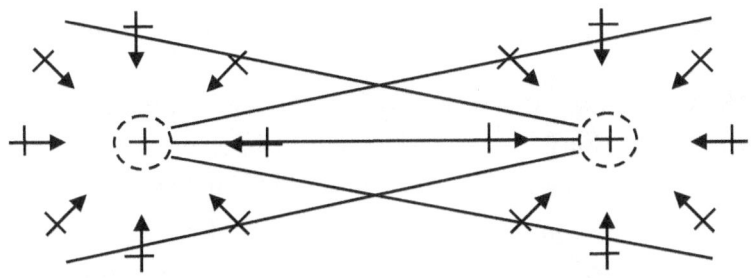

Fig. 6.2 Overlapping of charge lines

For two positrons, bit redistribution due to overlapping of charge lines results in more bits in the region between the two positrons, as shown in Fig. 6.3. The momentums of both positrons are changed such that the positrons are moving away from each other. It appears as if a repulsive force exists between the positrons. According to classical theory of electromagnetism, this is the electric force due to the electric field of the positrons. The situation of two electrons is similar to that of two positrons.

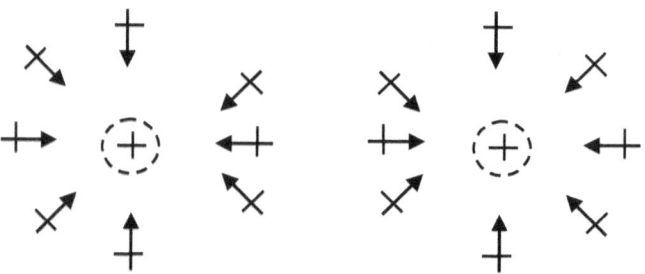

Fig. 6.3 Bit redistribution of two positrons

If one of the positrons in Fig. 6.3 is replaced with an electron, the charge lines of the two particles are of the opposite sign. The result of bit redistribution due to overlapping of charge lines is that there are fewer bits in the region between the two particles, as shown in figure 6.4. The momentums of the two particles are changed in such a way that the particles are moving toward each other. It appears as if an attractive force exists between the positron and the electron. In other words, the electric force is attractive.

Four-Element Model of Particle Physics

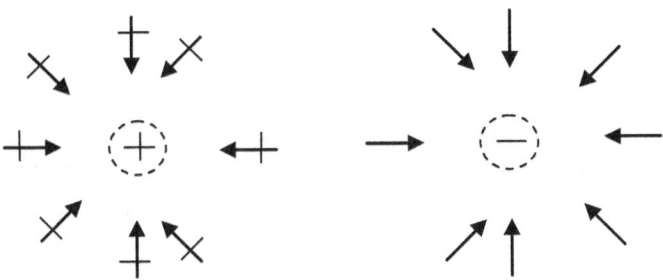

Fig. 6.4 Bit redistribution of an electron and a positron

A charge pair is formed because of the electric force between a positron and an electron. Once the charge pair is formed, the two charges stay together until being separated by a photon.

Magnetic Force

When a charge moves, it is the charge center that actually moves, and consequently the charge lines associated with the charge center will change. Thus, when a charge moves, it keeps switching charge lines. Figure 6.5 shows switching of charge lines when a charge moves from left to right. The charge lines actually fill 3-dimensional space, and Figure 6.5 is a 2-dimensional view only.

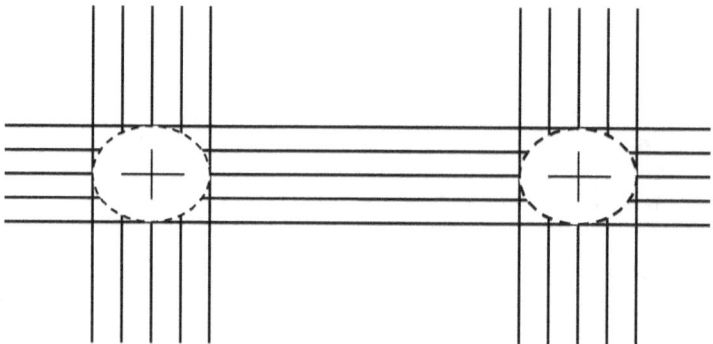

Fig. 6.5 Switching of charge lines

When two charged particles are both moving in a specific reference frame; switching of charge lines can result in change of the distribution of bits around both particles, and results in a force between the two particles. This force is the magnetic force in classical theory of electromagnetism. The magnetic force between two charged particles exists in a specific reference frame only when both particles are moving.

Emission of Bits

When a charged particle is at rest or moves with constant speed, its constituent bits will stick to the particle. However, when the particle changes its direction or speed, some bits can leave. In other words, when a particle changes its direction or speed, some bits can be emitted. This is like the situation in which a group of people pushing a cart together. When the cart moves with fixed direction and speed, everybody can hold on to the cart. But when the cart suddenly changes its speed or direction, some people may not be able to hold on. If the cart stops completely all in a sudden, everybody will rush forward and separate from the cart.

Electromagnetic Radiation

According to classical theory of electromagnetism, a moving electron or positron will emit electromagnetic radiation when its direction or speed changes. Since the electromagnetic radiation is a collection of photons, which contain n-bits and p-bits; the emission of electromagnetic radiation must be a joint work of the electron, which contains only n-bits, and the positron, which contains only p-bits.

An apparatus that is efficient for electromagnetic radiation is antenna. Figure 6.6 shows a simple antenna that consists of two conductors connected to a sinusoidal alternating-current (ac) power supply. The power supply causes an electric current to flow on each conductor. The amplitude of the current varies constantly, and the direction of the current changes periodically.

Four-Element Model of Particle Physics

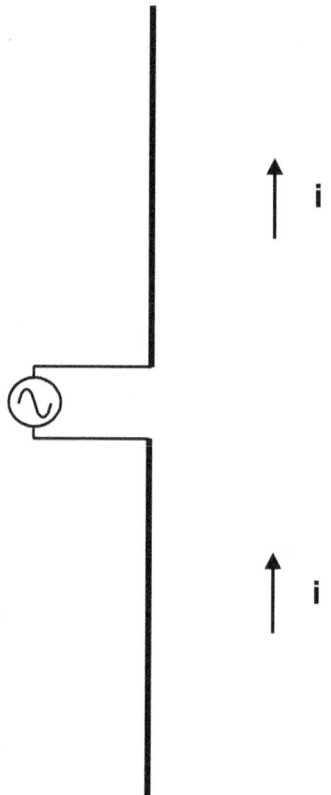

Fig. 6.6 A simple antenna

In a conductor, there are a lot of electrons that can move freely and there are a lot of positrons in atomic nuclei that can vibrate about their center position. The electric current in a conductor is the net result of motion of electrons and positrons.

Four-Element Model of Particle Physics

The motion of electrons and positrons is due to additional bits provided by the electric power supply. The situation is illustrated in Figure 6.7 in which the bits on the outer circle stand for additional bits obtained by an electron or positron.

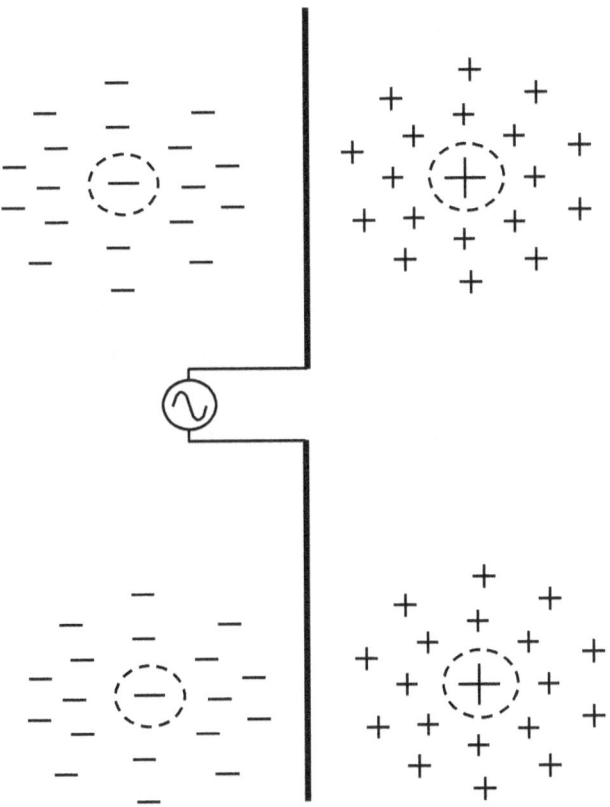

Fig. 6.7 Electron and positron with additional bits

Since the amplitude of the electric current varies constantly, the speed of the electrons and the vibration rate of the positrons vary constantly. As a result, some n-bits of an electron and some p-bits of a positron will leave the conductor together as a photon, as depicted in Figure 6.8. The number of photons emitted is proportional to the current flowing in the conductor, and the frequency of an individual photon is proportional to the frequency of the electric power supply.

An antenna is not only a good emitter of electromagnetic radiation but also a good receiver. When an electromagnetic wave emitted from an antenna encounters a receiving antenna, the bits of the photons will set the electrons and positrons in the conductors to motion. Electric currents thus flow in the conductors which are connected to an electronic devise such as the preamplifier of a radio.

Four-Element Model of Particle Physics

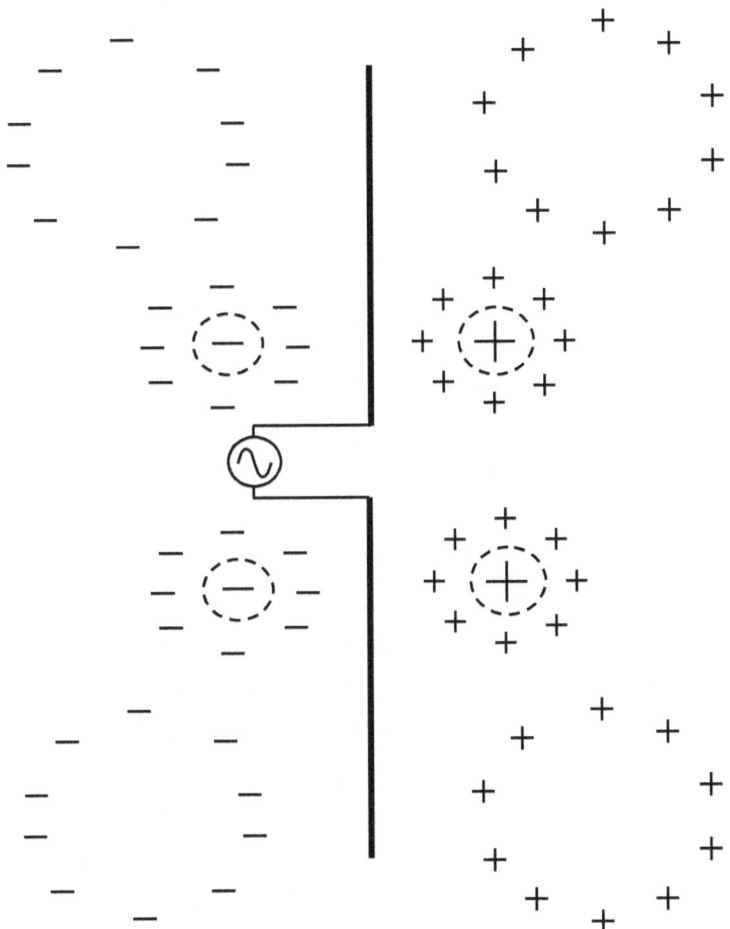

Fig. 6.8 Photon emission

Chapter 7: Origin of Mass and Gravity

What is the essence of mass? What is the origin of gravity? These are questions that are not well answered in classical physics. In this chapter, answers based on the four-element model will be provided.

Origin of Mass

Since the neutrino has zero or trivial mass, its constituents, the bits, must be the same. But when n-bits combine with a negative charge to form an electron, the electron has mass. This indicates that mass is the result of bits combing with charges, and that when bits combine with charges to form a particle, the energy of the bits is converted into the mass of that particle. When the bits are no longer bound to a particle, they turn back to the form of energy. This is in agreement with Einstein's famous equation

$$E = mc^2$$

which says that energy can be changed for mass, and vice versa.

When a negative charge attracts n-bits to form an electron, the energy of the n-bits is converted into the mass of the electron. The mass of an electron is thus in direct proportion to the number of n-bits in the electron. Similarly, the mass of a positron is in direct proportion to the number of p-bits in the positron. For a matter particle with both types of bits, the mass of the particle is in direct proportion to the total number of bits in the particle.

Since the rest mass of an electron is a constant, the number of bits of an electron at rest is fixed, which means that a charge at rest can attract only a certain number of bits. Thus, for a particle to have a rest mass larger than that of an electron, it must have more charges than an electron.

Heavy Particles

All heavy particles including quarks contain a large number of positive charges and negative charges, which in turn attract a large number of bits. The number of positive charges and the number of negative charges differ by the net charge unit of the particle. For example, in a neutron the number of negative charges is equal to that of positive charges, whereas in a proton the number of positive charges is larger than that of negative charges by one.

Positive charges and negative charges can form charge pairs. If this happens in a particle, the particle is not stable. Particles are stable because their constituent elements are distributed and move around so that charge pair formation cannot happen.

When a particle and its antiparticle meet, their constituent elements mix up, and the distribution and motion of the constituent elements change. As a result, every positive charge will pair with a negative charge to form a charge pair that cannot attract bits. The bits are thus detached from charges and form one or more photons.

Muon and Tauon

The muon differs from an electron only in that it has a much larger mass. Thus, the muon contains a large number of positive charges and negative charges. The number of negative charges is larger than that of positive charges by one. Because of the existence of both types of charges the muon contains both types of bits.

The muon decays into an electron, an anti-electron-neutrino, and a muon-neutrino. The process can be expressed as

$$\mu \rightarrow e^- + \bar{v}_e + v_\mu$$

In the decay process; one negative charge and some n-bits form an electron, and all remaining charges form charge pairs. The remaining n-bits become a muon-neutrino, and all p-bits become an anti-electron-neutrino.

The tauon has a structure similar to that of a muon except that it contains more charges and bits. It decays into an electron, an anti-electron-neutrino, and a tauon-neutrino.

Muon-neutrino and Tauon-neutrino

The electron-neutrino, the muon-neutrino and the tauon-neutrino are all formed by a group of n-bits. The only difference between them is the bit distribution.

Origin of Gravity

The gravitational force (gravity) and electromagnetic force are similar in many respects. They are both of infinite range and fall as the square of distance. Moreover, they have similar expressions for the strength.

The expressions for the gravitational force Fg and electromagnetic force Fem between two particles are shown below.

$$F_g = G\frac{m_1 m_2}{r^2}$$

$$G = 6.67 \times 10^{-11} \frac{N \cdot m^2}{(kg)^2}$$

$$F_{em} = k\frac{q_1 q_2}{r^2}$$

$$k = 9.0 \times 10^9 \frac{N \cdot m^2}{C^2}$$

In the expression for the gravitational force, G is a constant, m_1 and m_2 are the masses, and r is the distance. The force unit is in Newton (N), mass unit in kilogram (Kg), length unit in meter (m).

Four-Element Model of Particle Physics

In the expression for the electromagnetic force, k is a constant, q_1 and q_2 are electric charges, and r is the distance. The force unit is in Newton (N), the charge unit in Coulomb (C), the length unit in meter (m).

Similarity between the gravitational and electromagnetic forces is not a coincidence. The two forces are similar because they have the same origin.

According to the four-element model, there is only one type of force acting between two particles, and the force is resulted from interactions among constituent charges and bits of the two particles. The four fundamental forces in the standard model are manifestation of the same force under different circumstances. When the separation of the two particles is extremely small, say about or below the size of a nucleus, the force manifests as the strong force. When the separation is large, the force manifests as the electromagnetic and gravitational forces. When it comes to some particle transformations or decays, the force is called the weak force by particle physicists.

As mentioned above, the force between two particles is the result of interactions among all constituent charges and bits, and the magnitude of the force F is

$$F = F_{em} + F_g$$

Whereas F_{em} can be positive or negative depending on the sign of the electric charges, F_g is always positive. F_g is proportional to the mass of the particle because the mass is in direct proportion to the number of bits.

Chapter 8: Interaction between Photon and Atom

Interactions between photons and atoms are important subjects in atomic and particle physics. In this chapter, these interactions will be explained using the concept of the four-element model.

Interaction between Electron and Atomic Nucleus

An atomic nucleus contains protons and neutrons. The atomic nucleus is always electrically positive and thus can be regarded as an entity with positive electric charge.

An atom contains an atomic nucleus surrounded by one or more electrons. According to classical theory of electromagnetism, an electric field and magnetic field exist between the electrons and the nucleus. These fields keep the electron orbiting the nucleus as shown in Fig. 8.1.

Four-Element Model of Particle Physics

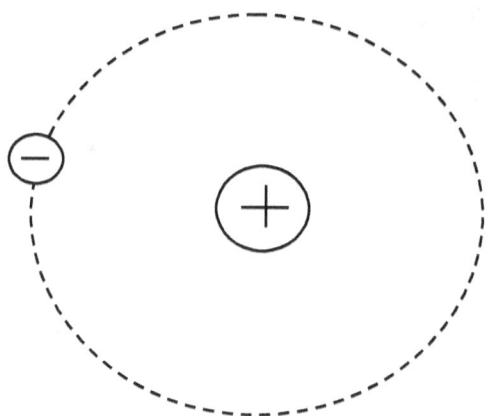

Fig. 8.1 Electron and atomic nucleus

From the viewpoint of the four-element model, it is the interactions between the constituent elements, i.e. charges and bits, of the electrons and those of the nucleus that keep the electrons moving around the nucleus.

Before the theory of quantum mechanics was developed, physicists had been puzzled by the fact that an electron moves around an atomic nucleus without losing energy. According to classical theory of electromagnetism, an electron circling an atomic nucleus will emit electromagnetic radiation and lose energy consequently. The electron will eventually spiral into the nucleus. From the viewpoint of the four-element model, a particle loses energy only when it loses some of its bits. Interactions between the constituent charges and bits of the electron and the nucleus result in a force that keeps the electron moving around the nucleus without losing any bit.

Interaction between Photon and Atom

An atom can be excited by absorbing a photon, as shown in Figure 8.2, and one of its electrons gains energy and jumps to an outer orbit.

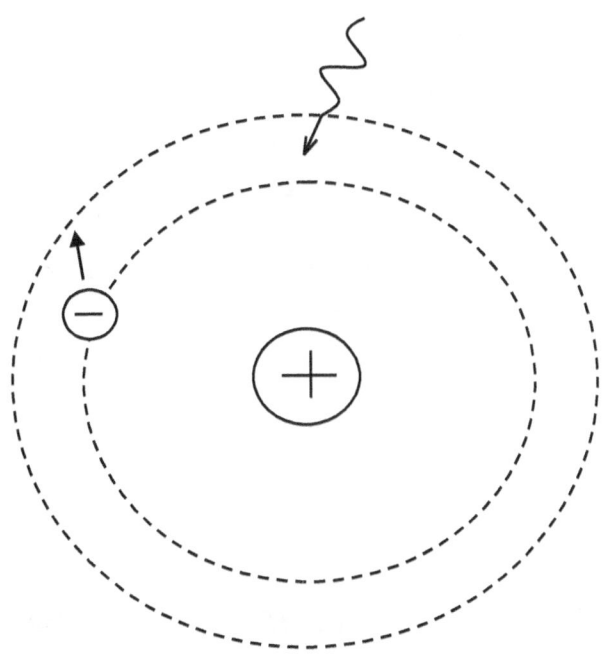

Fig. 8.2 Photon absorption by atom

According to conventional physics, the photon is absorbed by the electron, and all of the energy carried by the photon is transferred to the electron.

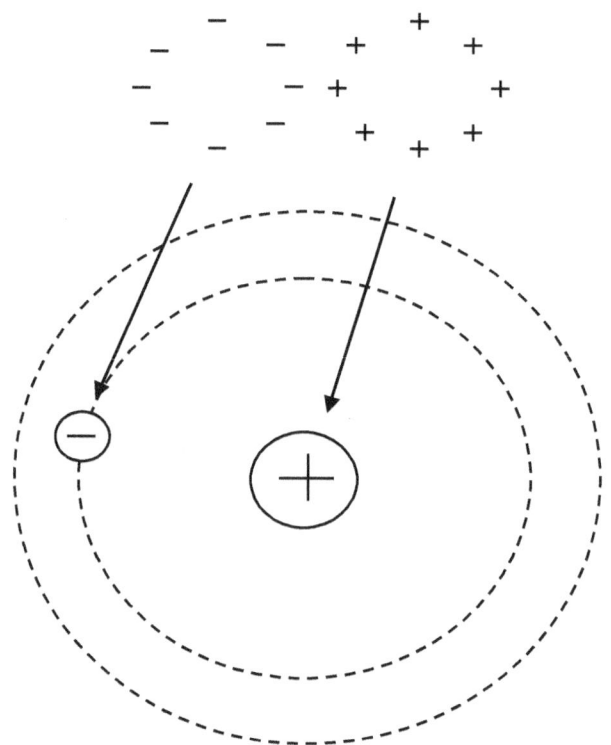

Fig. 8.3 Absorption of n-bits and p-bits

Since the electron contains only n-bits, it can absorb only n-bits of the photon. The p-bits of the photon must be absorbed by the nucleus as shown in Fig. 8.3. The question is: does the electron acquire an amount of energy equal to that originally carried by the photon? There are two possibilities. The first one is that the electron acquires only the energy carried by the n-bits of the photon, and the energy carried by the p-bits is acquired by the nucleus. The second one is that the nucleus absorbs the p-bits of the photon and gives away some n-bits to the electron. The electron thus acquires an amount of energy equal to that originally carried by the photon.

Photoelectric Effect

When light shines on metal surface, some electrons can acquire energy from the photons and escape from the surface, as shown in Fig. 8.4. This phenomenon is called photoelectric effect.

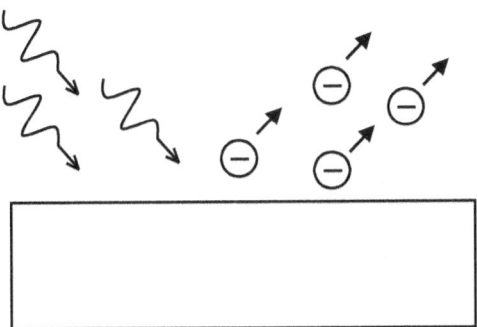

Fig. 8.4 Photoelectric Effect

It is known that photoelectric effect cannot happen to a free electron as energy and momentum cannot both be conserved. Thus, photoelectric effect must be an interaction between a photon and an electron bound to an atom. In other words, photoelectric effect is an interaction between a photon and an atom.

Compton Effect

The Compton Effect is a process in which a photon scatters from an electron as shown in Fig. 8.5. In the process, the photon transfers part of its energy to the electron, and as a result, its frequency decreases.

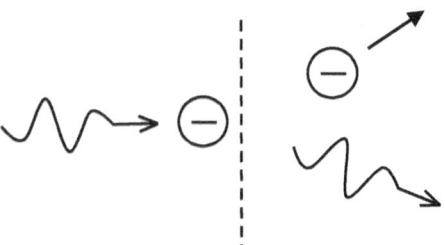

Fig. 8.5 Compton Effect

Does the photon give away only n-bit or both types of bit? If the photon gives away only n-bits, the Compton Effect is an interaction between a photon and an electron. If the photon gives away both n-bits and p-bits, the Compton Effect must be an interaction between a photon and an atom.

X-ray Production

When a high-energy electron passes by a nucleus, as shown in Fig. 8.6, it decelerates and emits an X-ray photon. The radiation produced in this way is often referred as bremsstrahlung (German for "braking radiation").

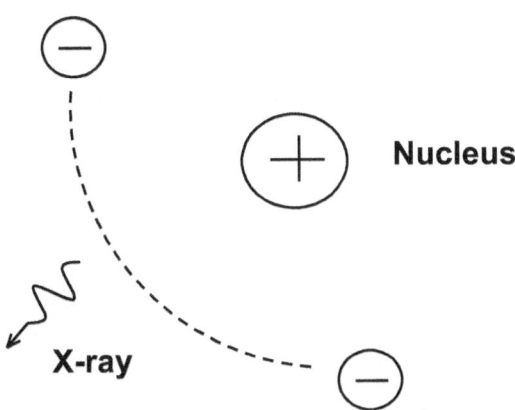

Fig. 8.6 X-ray production

Four-Element Model of Particle Physics

Since the electron contains only n-bits, the p-bits of the X-ray photon must come from the nucleus. In other words, the energy of the photon is partly from the electron, and partly from the nucleus.

Chapter 9: Quark Mystery

According to the standard model, the quark has two unusual characteristics. First, the quark has color charge which is the source of the strong force. Second, the quark is permanently confined inside hadrons because of a feature called asymptotic freedom. In this chapter, I am going to raise questions on color charges, and explain quark confinement without invoking the idea of asymptotic freedom.

Color Charge

When a high-energy positron and a high-energy electron collide, a pair of quark and anti-quark can be produced, as shown in Fig. 9.1, and the process can be expressed as

$$e^+ + e^- \rightarrow \gamma \rightarrow q + \bar{q}$$

Neither the positron nor the electron contains color charge. Where are the color charges of the quark and anti-quark from?

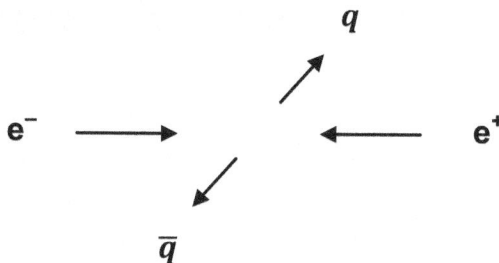

Fig. 9.1 Quark-antiquark pair creation

When a hadron and its antiparticle meet, they annihilate each other and become a flash of light (photons). The hadron is made from quarks, which contain color charges, but the photon does not contain color charge. Where do the color charges go?

The above cases led me to believe that color charges are not real. Different colors just represent different states of a quark, and each state has a specific distribution of the constituent elements.

Quark Confinement and Hadronization

According to the standard model, the force between two quarks is the color force due to the color charges of the quarks. This force increases as the separation between the two quarks increases. When an external energy is applied to separate two quarks from each other, the color force increases with the separation and the two quarks can never be separated from each other completely. Because of this strange feature quarks are bound together and confined permanently inside a hadron. This phenomenon is referred to as asymptotic freedom by particle physicists. When external energy exceeds a certain level, one of the quarks can transform from one type to another, and in addition, new hadrons can be produced, a phenomenon called hadronization by particle physicists.

From the viewpoint of the four-element model, quarks are, like other particles, formed by bits and charges. A quark maintains its specific structure by interacting strongly with other quark, and this is possible only inside a hadron. If a quark were able to escape from a hadron, it would decay into something else. When external energy, which is in the form of particles such as photons and electrons, is applied to a hadron; the constituent parts of the incoming particle mingle and interact with those of the quarks. The result is that the structures of the original quarks can change, and new quarks and new hadrons in turn can also be produced. Thus, quarks always reside in a hadron and cannot be observed as free quarks. Quark confinement needs not invoke the idea of color force and asymptotic freedom at all.

Chapter 10: Miscellaneous Topics

This chapter features miscellaneous topics including wave-particle duality, energy and momentum conservation, and neutrino mass.

Wave-particle duality

When a light beam passes through the slit of a screen and reaches a second screen, as shown in Fig. 10.1: the distribution and intensity on the second screen depends on both the width of the slit and the wavelength of the light beam. If the width is large compared to the wavelength, light distribution is as shown in Fig. 10.1. It appears that light travels along a straight line through the slit to the second screen.

Four-Element Model of Particle Physics

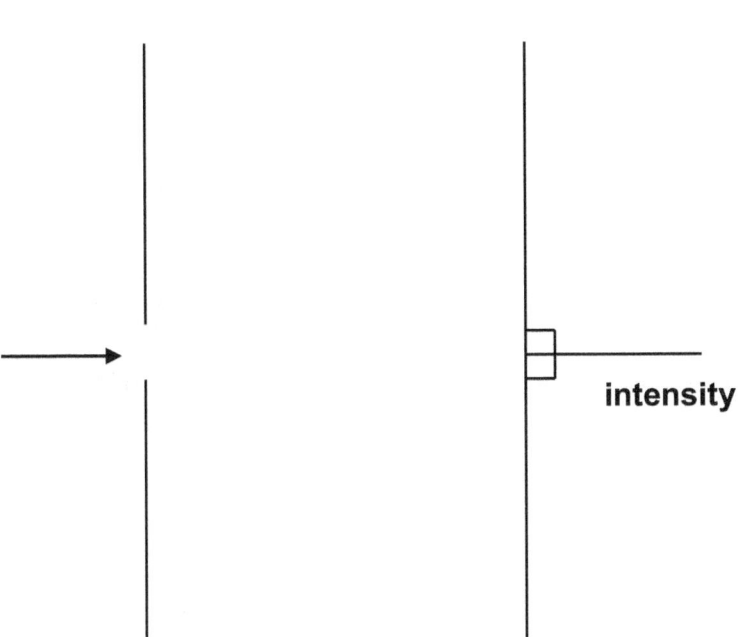

Fig. 10.1 Light without diffraction

If the width of the slit is comparable to the wavelength, light distribution is as shown in Fig. 10.2. It appears that some light turns around the corner of the slit, and reaches some area that cannot be reached if light behaves as a collection of particles. This phenomenon is called diffraction. Diffraction can be explained only by the wave property of light.

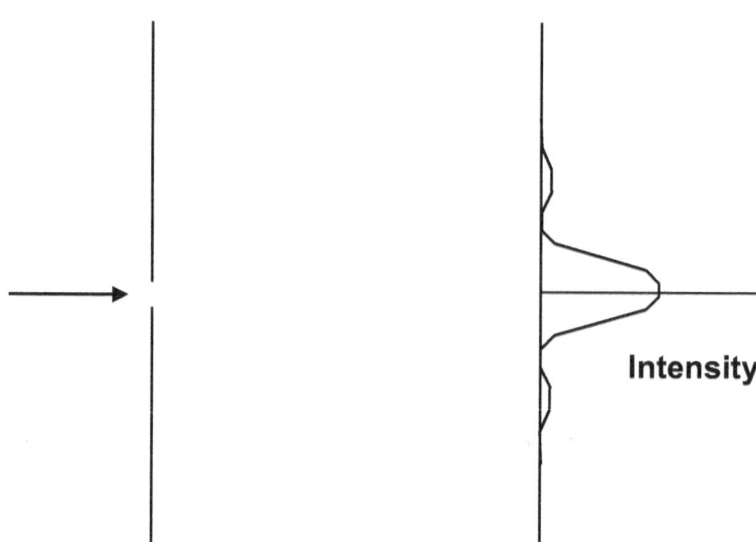

Fig. 10.2 Diffraction of light

When an additional screen with two slits is inserted between the two screens of the above experiment, and the widths of the slits are comparable to the wavelength of the light beam; the distribution and intensity of the light on the right screen is as shown in Fig. 10.3. The light distribution appears as alternating bright and dark bands. This phenomenon is called interference. A conventional explanation for interference is given below.

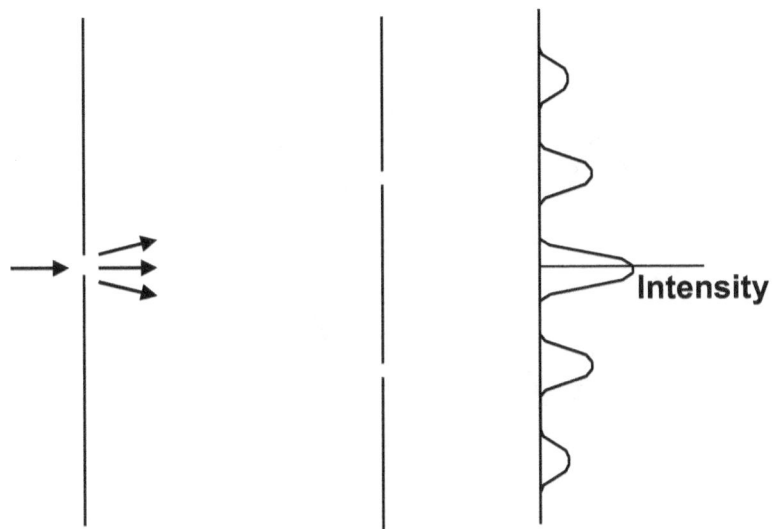

Fig. 10.3 Interference of light

The two slits on the middle screen allow the light to travel by two different paths. Light from the two paths can have different phase when they arrive at the same point on the right screen. When they are in phase completely, they reinforce each other and the intensity is doubled as shown in Figure 10.4. When they are out of phase completely, they cancel each other and the intensity becomes zero as shown in Figure 10.5. For cases in between, they can either reinforce or cancel each other depending on the phase difference.

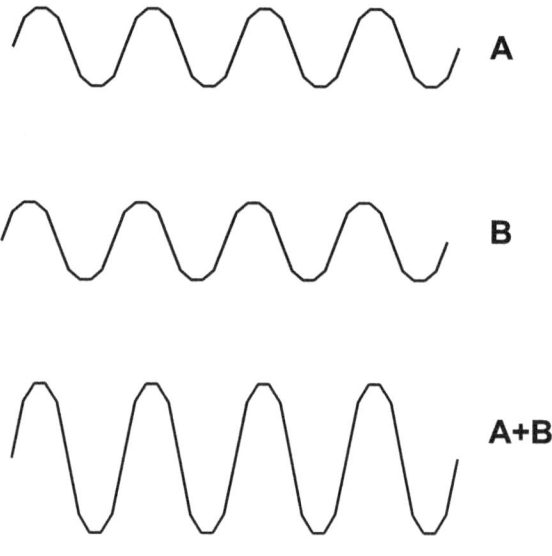

Fig. 10.4 Total reinforcement

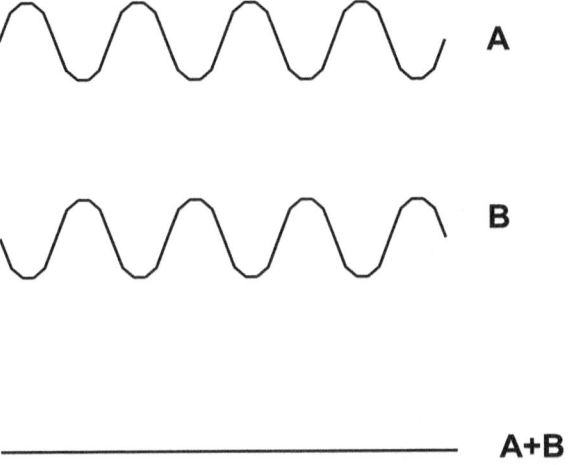

Fig. 10.5 Total cancellation

If light is regarded as a collection of particles, light interference can be interpreted as interference between individual photons. However, when the light intensity is reduced such that at each time instant there is only one photon, the photon has no other photon to interfere with, but interference still occurs in all experiments of this type. One possibility is that a photon can split into two parts, and each of them passes through one of the two slits of the middle screen. The two parts then interfere with each other on the right screen. But experiments have demonstrated that the photon passes only one of the slits, not both.

Four-Element Model of Particle Physics

Is light a wave or a collection of particles? Both classical physics and quantum physics get around this with the idea of wave-particle duality. Light manifests as a wave under certain circumstances and particles under other circumstances.

According to the four-element model, particles consist of charges and bits, and each constituent bit is rotating about a center of its own. Such a structure cannot be described simply as a particle or as a wave. A more important point is that the photon does not interact with other photons. Since an electromagnetic wave is a collection of photons, it cannot interact with other electromagnetic waves. Thus, light interference is not the result of two light waves interfering with each other.

When a photon goes through the slit of the screen, it can interact with atoms of the screen. As a result, the photon is deflected as shown in Figure 10.6. The amount of deflection is determined by the frequency, or equivalently the energy, of the photon and the dimension of the slit. Thus, photons will arrive at different locations of the right screen with different probability; the intensity of light on the right screen varies and this results in diffraction in the one-slit experiment and interference in the two-slit experiment.

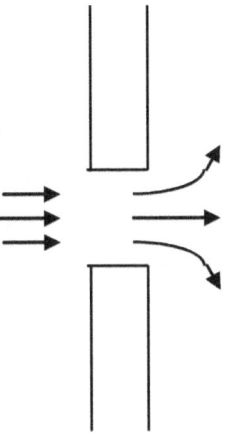

Fig. 10.6 Light deflection at the slit

Energy and Momentum Conservation

The total energy, which is the sum of the rest mass and kinetic energy, of a particle is determined by the number of bits the particle has. Since bits cannot be created and cannot be destroyed, the initial and final total number of bits must be the same in all particle interactions. In other words, energy is conserved in all particle interactions without exception.

Four-Element Model of Particle Physics

In an isolated system, i.e. a group of particles interacting with one another without external influence, the momentum change of a particle is always accompanied by momentum changes of other particles. As a result, the momentum of the system in conserved. This can be illustrated by a system consisting of two positrons as shown in Fig. 10.7. The two positrons are originally at rest, and each has zero momentum. The positrons then move away from each other because of the electric force. Now they have momentums that are equal in magnitude but in opposite directions. The net momentum of the system is still zero.

Four-Element Model of Particle Physics

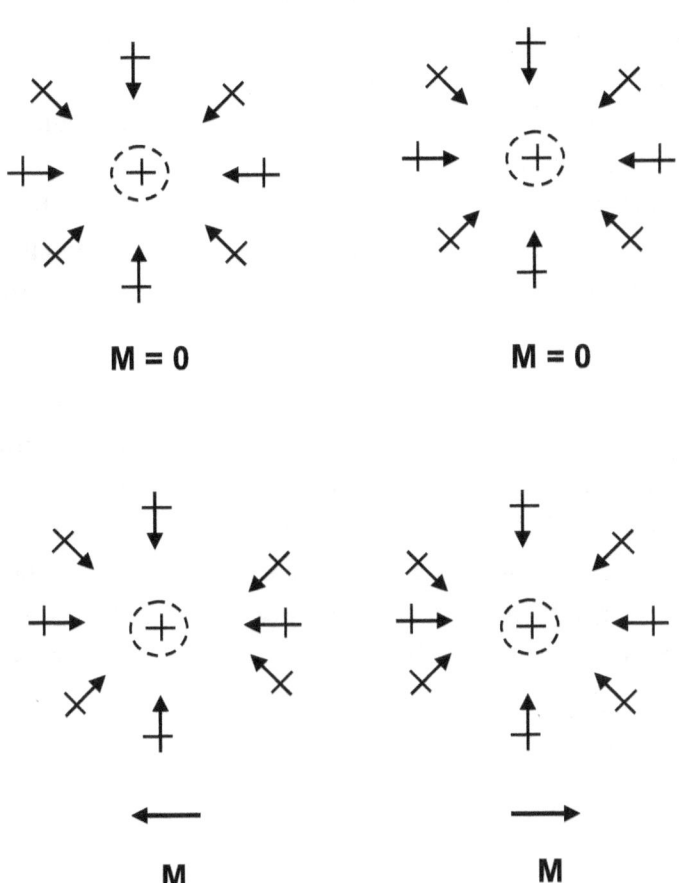

Fig. 10.7 Momentum conservation

Neutrino Mass

Mass is the result of combining charges and bits. Since the neutrino has no charge, it should have zero mass. However, the constituent bits of a neutrino can interact with constituent charges of matter particles. Since the neutrino has no charge, this kind of interaction is not of the electromagnetic nature. The only way to account for this in conventional physics is to say that the neutrino has mass. However, the mass of a neutrino is by nature different from that of a matter particle and requires a new theory for predicting its magnitude.

Photons are, like neutrinos, formed by bits. As the neutrino has mass, the photon too should have mass. But the consensus is that the photon has zero mass. However, according to Einstein's general theory of relativity, a light ray can be deflected by a body of heavy mass such as a planet. This means that light does respond to mass (gravity). In other words, the photon may have mass.

References

Elementary Modern Physics, Third Edition; Richard T. Weldner and Robert L. Sells, (Allyn and Bacon, Inc., Boston, 1980)

Particle Physics: Brian R. Martin, (Oneworld Publications, London, 2014)

Particle Physics – A Very Short Introduction: Frank Close, (Oxford University Press, Oxford, 2004)

The Particle Garden; Gordon Kane, (Helix Books, 1996)

Facts and Mysteries in Elementary Particle Physics; Martinus Veltman, (World Scientific, 2003)

Quarks and Gluons – A Century of Particle Charges; M Y Han, (World Scientific, 1999)

The Wizard of Quarks; Robert Gilmore, (Copernicus Books, New York, 2001)

Science or Fiction ? The Phony Side of Particle Physics; Ofer Comay, (Samuel Wachtman's Sons Inc., Monterey, California, 2014)

From Quarks to the Cosmos; Leon M Lederman and David N Schramm, (Science American Library, New York, 1995)

The Grand Design; Stephen Hawking and Leonard Mlodinow, (Bantam Book, New York, 2010)

Neutrino; Frank Close, (Oxford University Press, New York, 2010)

Quantum Physics: Illusion or Reality ? ; Alastair Rae, (Cambridge University Press, 1996)

QED; Richard P. Feynman, (Princeton University Press, Princeton and Oxford, 2006)

The Hunting of the Quark; Michael Riordan, (Simon & Shuster, Inc., New York, 1987)

Student Friendly Quantum Field Theory, Second Edition; Robert D. Klauber, (Sandtrove Press, Fairfield, Iowa, 2013)

Quantum Physics of Atoms, Molecules, Solids, Nuclei, and Particles , Second Edition ; Robert Eisberg and Robert Resnick, (John Wiley & Sons, 1985)

Inside the nucleus; Irving Adler, (The John Day Company, New York, 1963)

Basic Concepts in Relativity; Robert Resnick and David Halliday, (MacMillan Publishing Company, New York, 1992)

www.ingramcontent.com/pod-product-compliance
Lightning Source LLC
Chambersburg PA
CBHW020606220526
45463CB00006B/2468